BACKYARD BUGS

I SEE MOTHS

by Genevieve Nilsen

TABLE OF CONTENTS

tadpole
books

WORDS TO KNOW

brown

green

hair

orange

red

spots

I SEE MOTHS

hair

I see a moth!
It has hair.

This one is brown.

This one is green.

spot

This one has spots.

This one has orange.

This one has red.

This one looks
like a leaf.

Can you spot it?

LET'S REVIEW!

Moths can be many different colors. How does this moth blend in?

INDEX